They're coming back... to be colorful!

I0490223

SUPER DINOSAURS

SUPER DINOSAURS

Baby Rodas

SUPER DINOSAURS

This book is by:

_____

For coloring this book is preferable to use
colored pencils. In case you use felt-tip pens
the back of the page has been left blank
to avoid ruining your children's masterpieces

SUPER DINOSAURS

SUPER
DINOSAURS

Connect the dots!

DEINONYCHUS

SUPER DINOSAURS

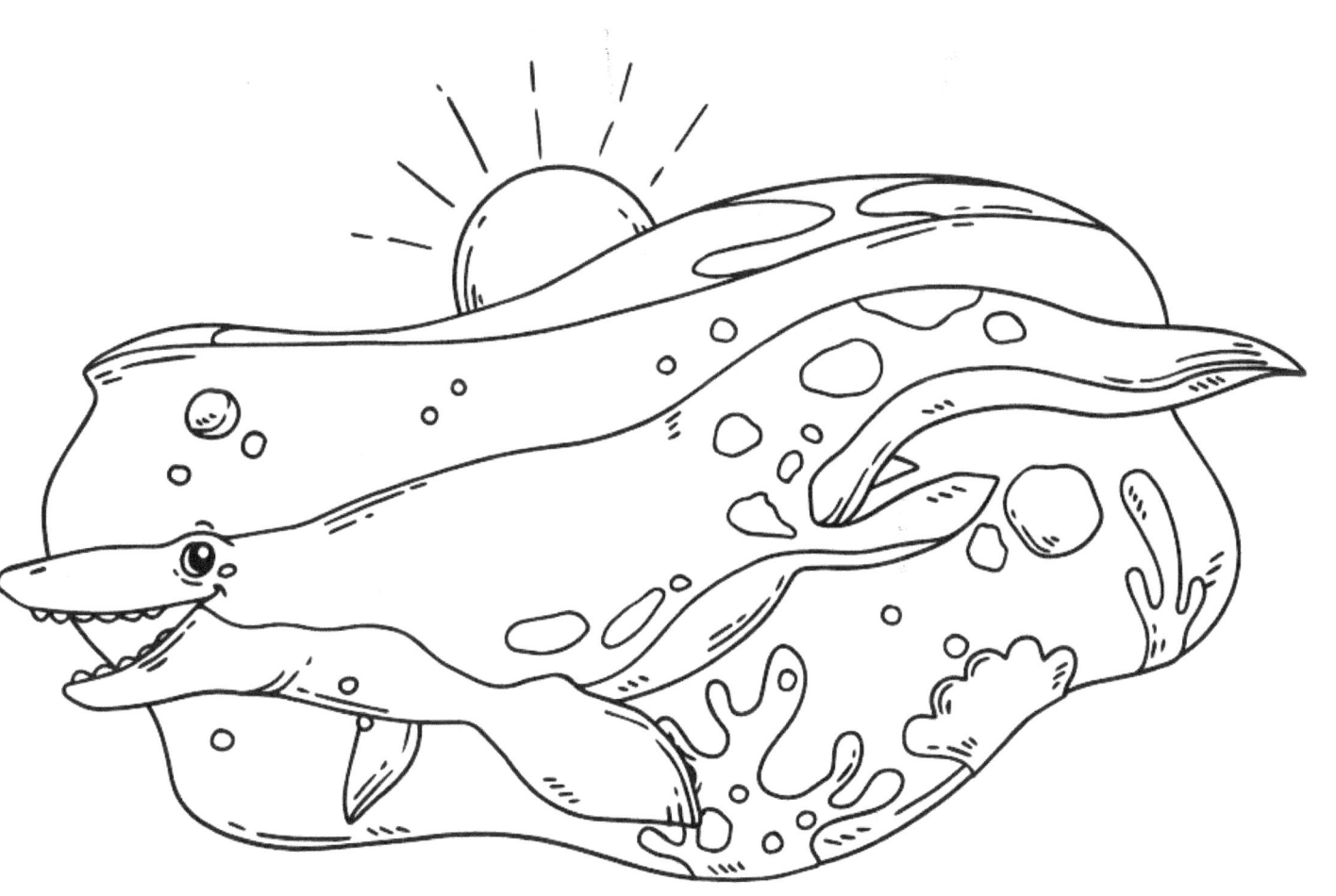

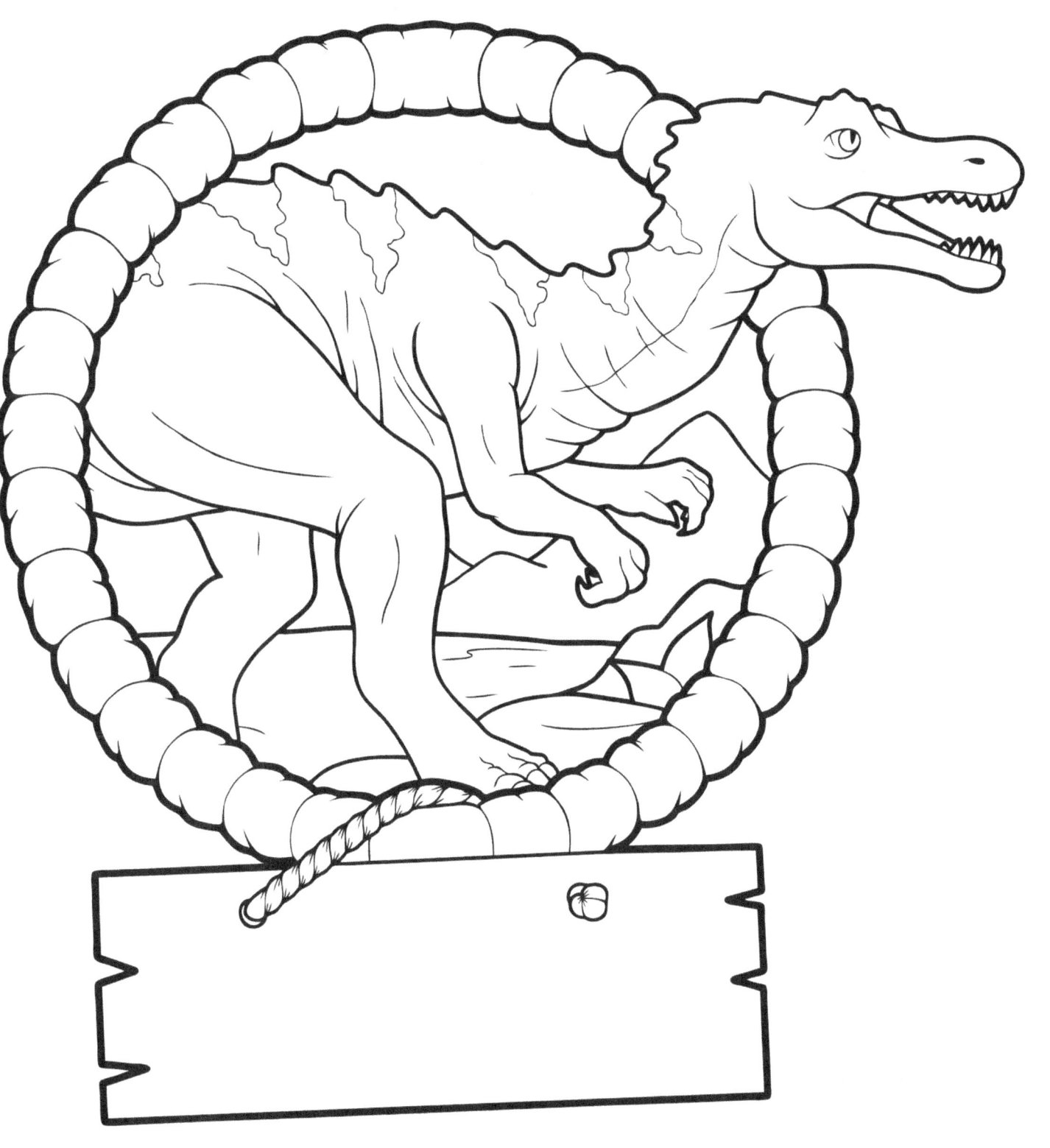

OVIRAPTOR

IGUANODON

www.ingramcontent.com/pod-product-compliance
Lightning Source LLC
Chambersburg PA
CBHW082147230526
45467CB00043B/2379